Collins

# Greater Depth Maths

Build curiosity and confidence

PUPIL RESOURCE PACK UPPER KEY STAGE 2



Herts for Learning

William Collins' dream of knowledge for all began with the publication of his first book in 1819. A self-educated mill worker, he not only enriched millions of lives, but also founded a flourishing publishing house. Today, staying true to this spirit, Collins books are packed with inspiration, innovation and practical expertise. They place you at the centre of a world of possibility and give you exactly what you need to explore it.

Collins. Freedom to teach.

Published by Collins
An imprint of HarperCollins*Publishers*
The News Building
1 London Bridge Street
London
SE1 9GF

HarperCollins*Publishers*
1st Floor, Watermarque Building
Ringsend Road
Dublin 4
Ireland

Browse the complete Collins catalogue at

**www.collins.co.uk**



10 9 8 7 6 5 4 3 2 1

ISBN 978-0-00-845492-0



British Library Cataloguing-in-Publication Data
A catalogue record for this publication is available from the British Library.

Authors: Charlie Harber, Rachael Brown, Nicola Adams, Laura Dell
Publisher: Lizzie Catford
Product manager: Letitia Luff
Editorial management: Oriel Square
Cover designer: Amparo Barrera
Cover photograph: Sasirin Pamai/Shutterstock
Typesetter: Jouve India Pvt. Ltd.
Internal illustrations: Jouve India Pvt. Ltd.
Production controller: Alhady Ali
Printed and bound in the UK using 100% Renewable Electricity at CPI Group (UK) Ltd

FSC
www.fsc.org
MIX
Paper from responsible sources
FSC™ C007454

This book is produced from independently certified FSC™ paper to ensure responsible forest management.

For more information visit: www.harpercollins.co.uk/green

The publishers gratefully acknowledge the permission granted to reproduce the copyright material in this book. Every effort has been made to trace copyright holders and to obtain their permission for the use of copyright material. The publishers will gladly receive any information enabling them to rectify any error or omission at the first opportunity.

Digital downloads available at:

**www.collins.co.uk/HertsforLearning/downloads**

# Contents

# The Hidden Number

## Curious

| | | | | | | |
|---|---|---|---|---|---|---|
| ? | ? | ? | ? | ? | ? | ? |

| | | | | | | |
|---|---|---|---|---|---|---|
| 0 | 1 | 2 | 3 | 4 | 5 | 6 |

← This is the hidden seven-digit number.

What do you notice?
What questions do you have?

**1** I can see …
I have noticed …

**2** I wonder how …
I wonder if ...
I wonder what …
I wonder how many …
I wonder what would happen if …

| | |
|---|---|
| **digits** | **number** |
| **consecutive** | **solutions** |
| **more / fewer** | **simpler** |

**3** My question to investigate is …

## Curious

## Express Pathway

How many hidden digits can you work out?

Fill in the boxes in the top grid to reveal the hidden seven-digit number.

**The digit in each white box must match how many times the digit in the grey box directly below it occurs in the complete hidden number.**

- The digit above the 0 must match the number of 0s in the hidden number.
- The digit above the 1 must match the number of 1s in the hidden number.
- The digit above the 2 must match the number of 2s in the hidden number.
- And so on.

*For example:*

*This digit must match the number of 0s in the hidden number. So if you write '4' in this box, you must then include four 0s in the hidden number.*

*For example:*

*This digit must match the number of 5s in the hidden number. So if you write '1' in this box, you must then include one 5 in the hidden number.*

| | | | | | | |
|---|---|---|---|---|---|---|
| ? | ? | ? | ? | ? | ? | ? |
| 0 | 1 | 2 | 3 | 4 | 5 | 6 |

| | |
|---|---|
| digits | number |
| consecutive | solutions |
| more / fewer | simpler |

# Connect

## 1. Solve a simpler but related problem

| | | | |
|---|---|---|---|
| | | | |
| 0 | 1 | 2 | 3 |

| | | | | |
|---|---|---|---|---|
| | | | | |
| 0 | 1 | 2 | 3 | 4 |

Do you notice anything about the solutions?

Do they have anything in common?

How might that help you solve the seven-digit puzzle?

## 2. Try then improve

How could you adapt your first attempt?

Are there any digits that you are certain of? Why?

How will you keep track of your attempts?

This combination of digits doesn’t work, so
I will …
I am certain that this digit will be in the number
because …

## 3. Make a model / act it out

What concrete resources will be most useful for this puzzle?

## 4. Make a table / spot a pattern

List the solutions in a table and compare them.

Can you spot any patterns?

Can you continue the patterns to find the eight-digit hidden number?

| | 0 | 1 | 2 | 3 | 4 | 5 | 6 | 7 | 8 | 9 |
|---|---|---|---|---|---|---|---|---|---|---|
| 4-digit hidden number | | | | | | | | | | |
| 5-digit hidden number | | | | | | | | | | |
| 6-digit hidden number | *No solution* | | | | | | | | | |
| 7-digit hidden number | | | | | | | | | | |
| 8-digit hidden number | | | | | | | | | | |

I have noticed that …
This leads me to predict …

## Collaborate

Try solving the puzzle on your own first.

Then share your ideas with a partner.

What have you discovered?

## Communicate

Write instructions for finding the hidden number.

Where did you start? What did you do next?

How did you know a particular digit would not work?

**Finding the hidden number**

*I wonder how you would solve the following puzzle.*

*Where would you begin?*

*You may like to have a go yourself first before reading on.*

| | | |
|---|---|---|
| **digits** | **strategies** | **place value** |
| **solutions** | **If … then …** | **consequently** |
| **As a result, …** | **therefore** | **For this reason,** |
| **My prediction was based upon …** | **However,** | **Due to the fact** |

## Create

Were there any questions asked at the start of the task that you now want to investigate?

Is there anything new that you are wondering now?

| | | |
|---|---|---|
| **additional** | **digit** | **order** |

## Consider

The trickiest part of the puzzle was … I think that this was because …

I was frustrated by … I think that was because … To overcome this, I …

When I found a combination of digits that satisfied all of the conditions, I felt … I think that this was because …

I needed to motivate myself to work on this task because … As a result, I …

Next time I tackle a task that frustrates me, I will …

# An Ancient Number System

## Curious

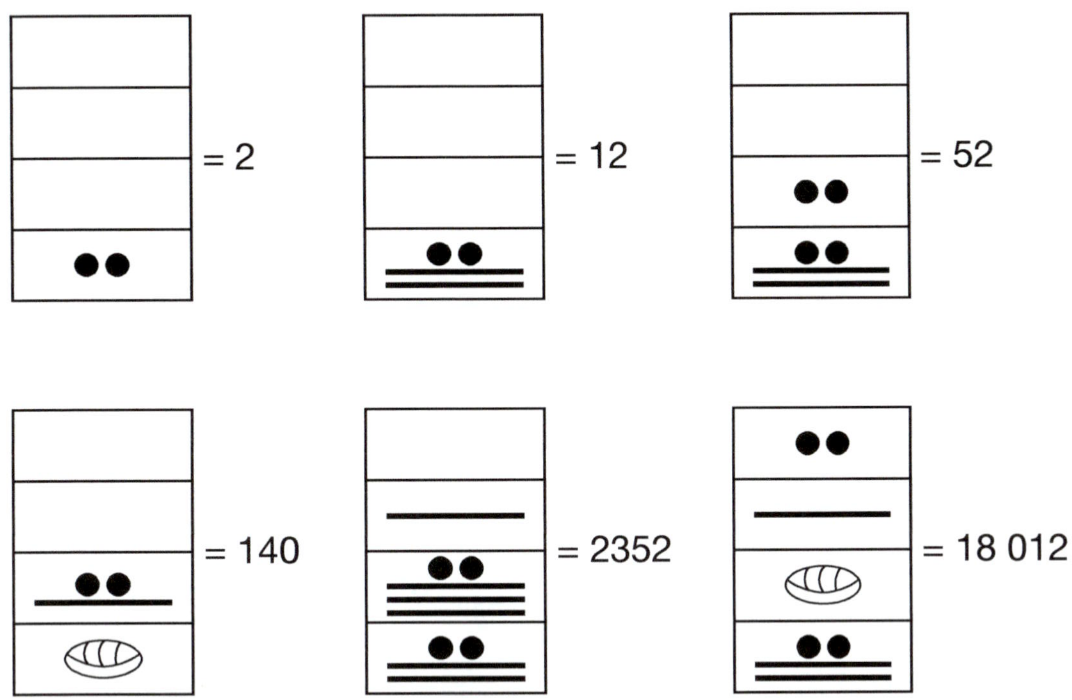


What do you notice?

What questions do you have?

**1** I can see …
I have noticed …
It reminds me of …

**2** I wonder where …
I wonder how …
I wonder who ….
I wonder if ….
I wonder what …

| | |
|---|---|
| **representations** | **base-10** |
| **number system** | **other** |
| **same** | **different** |
| **calculations** | **the ancient Maya** |

**3** My question to investigate is …

**Curious**

## Express Pathway

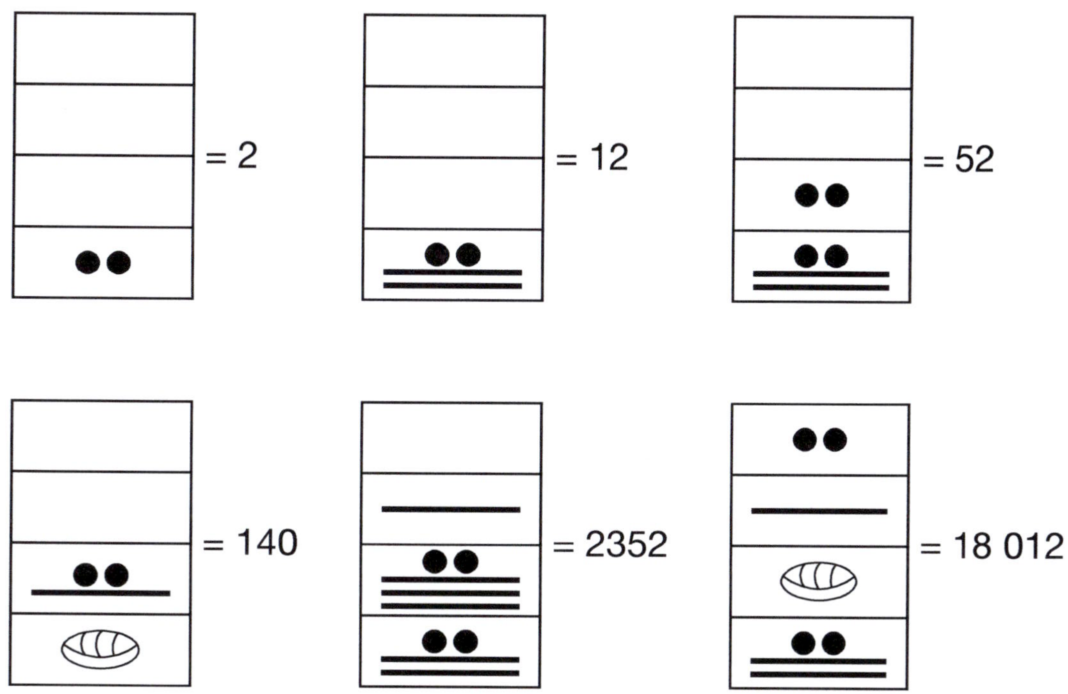


This is how the ancient Maya (a civilisation which dates back to around 1100 BCE in Central America) recorded numbers.

Can you work out how their number system works?

| | |
|---|---|
| representations | base-10 |
| number system | other |
| same | different |
| calculations | the ancient Maya |

# Connect

## 1. Make a model / act it out

Use counters and a place-value grid to represent the same number in each number system.
What's the same? What's different?

What row headings will you use for the Maya place-value grid?

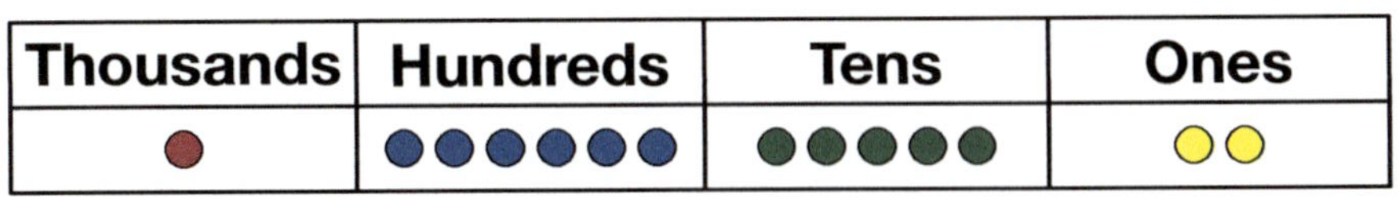


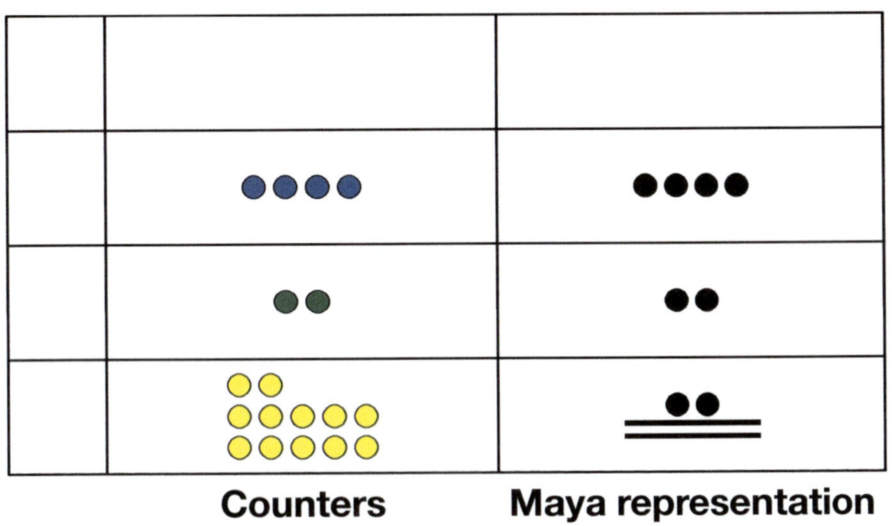


## 2. Make a table / spot a pattern

Make a table to deconstruct each representation.

Include a key.

**Key**

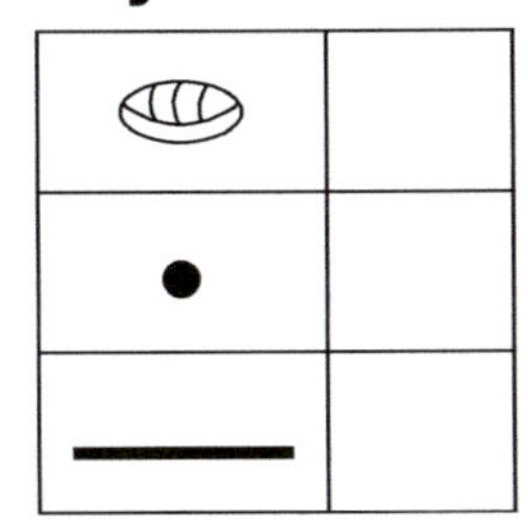

| **Whole number** | **2** | **12** | **52** |
|---|---|---|---|
| **Number of 8000s** | 0 | 0 | 0 |
| **Number of 400s** | 0 | 0 | 0 |
| **Number of 20s** | 0 | 0 | 2 |
| **Number of 1s** | 2 | 12 | 12 |
| **Maya representation** | 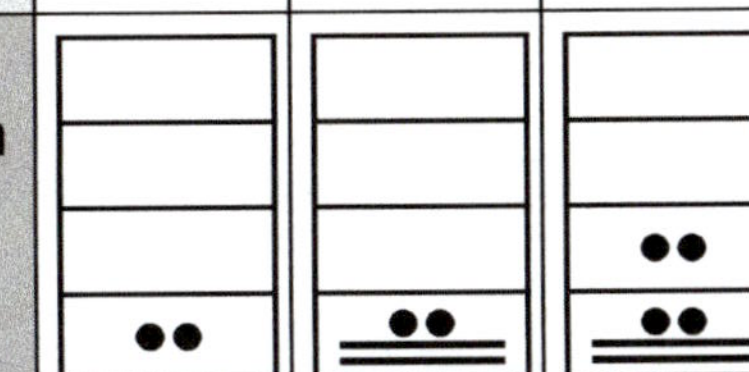 | | |

## 3. Draw a diagram

What would be useful to draw when translating from Maya representations to our numerals?

## Collaborate

**Game 1**
Roll a 0–9 dice three times to create a three-digit number.
Each represent the number in the Maya form.
Score a point if you are both correct.

**Game 2**
Sit back-to-back with your partner.
Describe your representation.
Your partner draws it and translates the number.
Did they get it right?

## Communicate

You are an explorer. You have found markings in an ancient Maya temple that you think represent numbers. Write a journal entry. Explain how the markings form a number system and how it differs from ours.

May 1842

*Today we continued exploring the ruins. On entering the first temple, once our eyes had grown used to the light, we saw …*

| | | |
|---|---|---|
| **Maya** | **ancient** | **number** |
| **system** | **symbols** | **Once we …** |
| **We ascertained that …** | **similar / dissimilar to** | **We deduced that …** |
| **We believe that …** | **pebbles, sticks and shells** | **representations** |

## Create

Were there any questions asked at the start of the task that you now want to investigate?
Is there anything new that you are wondering now?

| | | |
|---|---|---|
| **larger** | **ancient number systems** | **my own** |

## Consider

Others in the classroom had different ideas to me.
For example, …
I said 'aha' when … I said it because …
I was surprised about … because …
During this task, I saw myself as a thinker because …

# The Magic Circle

## Curious

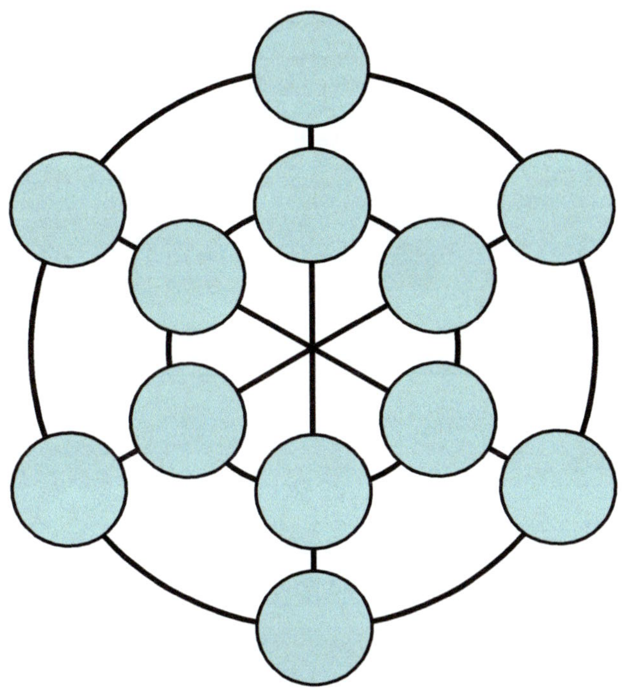

This is a magic circle.

What do you notice?

What questions do you have?

**1** I can see …
I have noticed …
It reminds me of …

**2** I wonder what …
I wonder whether …
I wonder how …
I wonder if …

| | | |
|---|---|---|
| **add** | **total** | **numbers** |
| **same** | **inner** | **outer** |
| **rings** | **solutions** | **sum** |

**3** My question to investigate is …

Curious 

Express Pathway

This is a magic circle.

The inner and outer rings both have the same total.

The vertical and diagonal lines all have the same totals.

Using the numbers 1–12, can you complete this magic circle?

Is there more than one solution?

| | | |
|---|---|---|
| add | total | numbers |
| same | inner | outer |
| rings | solutions | sum |

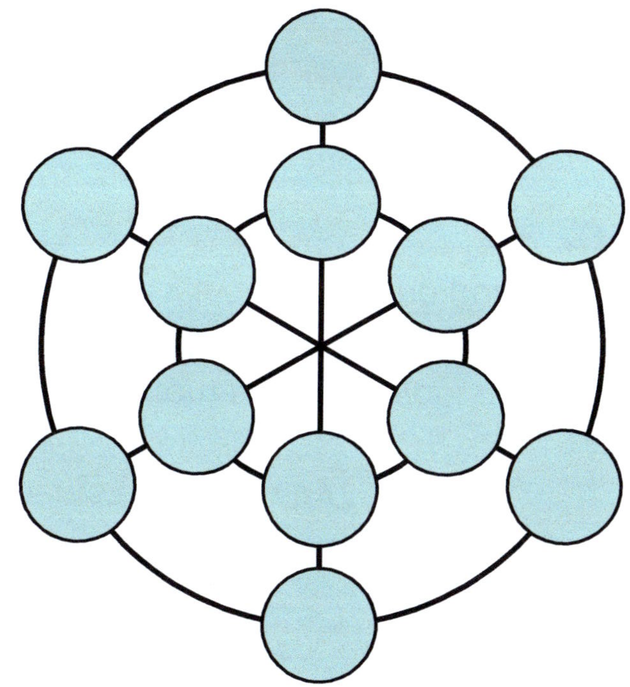

## Connect

### 1. Solve a simpler but related problem

Can you solve this smaller magic circle using the digits 1–8?

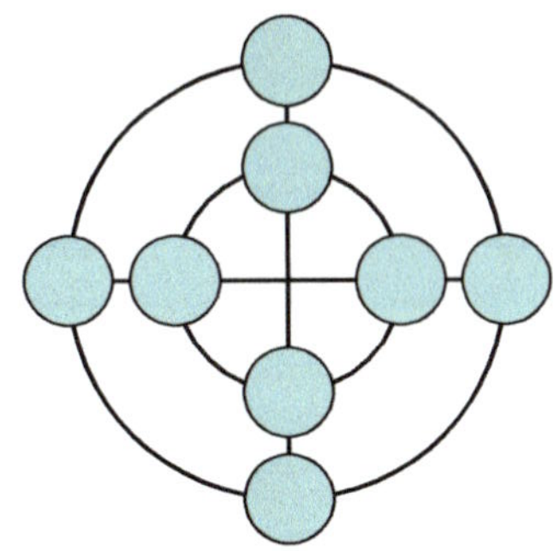

### 2. Work backwards / draw a diagram / make a model / act it out

Start with the total of the digits and then work backwards.

The total of all the numbers is …
If … equals …, then …
The … must total …
Now that I know this, I also know …

How might a bar model help?

How might number rods help?

### 3. Try then improve

What arrangements could you try?

Will you need to start from scratch if your arrangement does not work?

### 4. Create a systematic list

What arrangements could you try?
Write them in a list.

Will you need to start from scratch if your arrangement does not work?

Is there more than one way to solve the puzzle?

## Collaborate

Be prepared to talk about the strategies you used with another group.

What will you want to find out about how others approached the puzzle?

What interesting questions will you ask?

## Communicate

Write an explanation of how a magic circle works and how you could solve or design your own.

Make sure each step of your explanation is in the right order.

Address your reader directly.

> **Magic circles – an explanation of the phenomenon!**
>
> *You may have heard of a magic square, but have you ever heard of a magic circle?*
>
> *No? "What is a magic circle," I hear you ask?*

| | | | |
|---|---|---|---|
| **First, …** | **Begin by …** | **After that, …** | **When …** |
| **At this point, …** | **If …, then …** | **Consequently, …** | **If this works, then …** |
| **It might be advisable to …** | **The final stage will involve …** | **You'll be surprised to know that …** | **Now that you know more about …** |
| **total** | **sum** | **divide** | **inner and outer rings** |

## Create

Were there any questions asked at the start of the task that you now want to investigate?

Is there anything new that you are wondering now?

| | | | |
|---|---|---|---|
| **ring** | **create** | **line** | **consecutive numbers** |

## Consider

The trickiest part of solving the puzzle was …
The most important mathematics I learned by working on this puzzle was …
Working with others on this puzzle made me feel …
I interacted with others by …

# Does It All Add Up?

## Curious

This is an addition pyramid.

Each box is the sum of the two boxes directly below it.

Place the values below in the bottom row of the pyramid.

Can you find different ways to complete the pyramid?

| | | | |
|---|---|---|---|
| 0·6 | 0·7 | 0·9 | 1·5 |

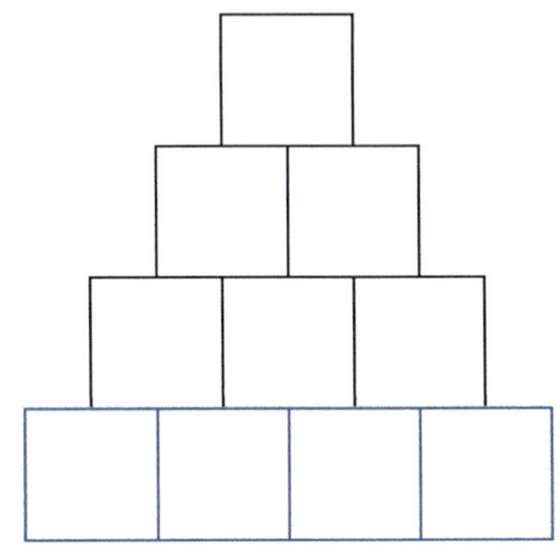

What do you notice?

What questions do you have?

**1** I can see …
I have noticed …

**2** I wonder if …
I wonder how …
I wonder what …
I wonder what would happen if …

| | | | |
|---|---|---|---|
| **possible** | **total** | **smallest** | **closest** |
| **sum** | **combinations** | **largest** | **arrangement** |

**3** My question to investigate is …

Curious

## Express Pathway

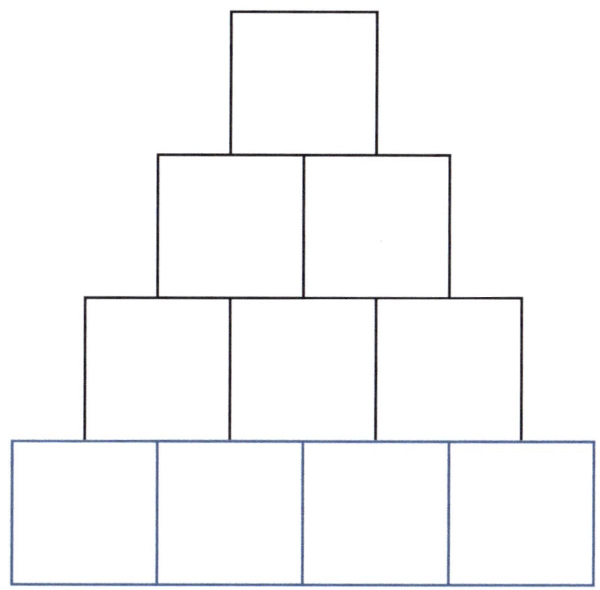

This is an addition pyramid.

Each box is the sum of the two boxes directly below it.

Arrange these values in the bottom row of the pyramid to create the smallest possible value for the top box.

| | | | |
|---|---|---|---|
| 0·6 | 0·7 | 0·9 | 1·5 |

| | | | |
|---|---|---|---|
| possible | total | smallest | closest |
| sum | combinations | largest | arrangement |

## Connect

### 1. Solve a simpler but related problem

Use the numbers 1, 2 and 3 in the bottom row of this pyramid.

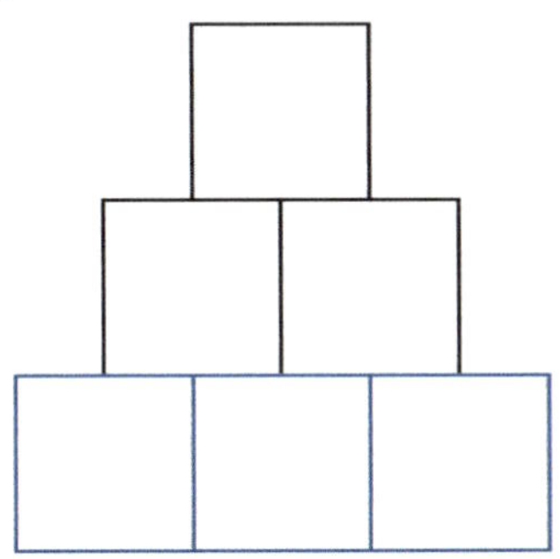

| smallest total | largest total | repeated total |
|---|---|---|

How do you know that you have tried all arrangements?

### 2. Draw a diagram

Use coloured shapes to expose and explain the structure of the pyramid.

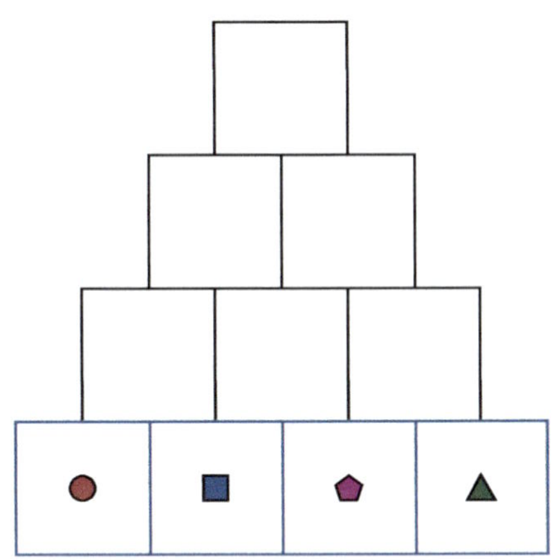

Could you use something else other than coloured shapes?

### 3. Create a systematic list / make a table / spot a pattern

How could you keep track of each arrangement and the total?

How will you organise this systematically?

| Numbers along the bottom row (left to right) | | | | Total at the top of the pyramid |
|---|---|---|---|---|
| 0·6 | 0·7 | 0·9 | 1·5 | 6·9 |
| | | | | |

## Collaborate

Complete a strategy map to compare your strategy with other people's.

What do you notice?

| | My strategy | Alternative strategy | Alternative strategy |
|---|---|---|---|
| Show it! | | | |
| Explain it! | | | |
| Critique it! | | | |

## Communicate

Compare and contrast your strategies with someone else.

Share your models, jottings and notes in order to justify your chosen approach.

Listen carefully to each other's explanations and ask for clarification if necessary.

I'm sorry, but I don't quite understand.
Could you show me in a different way?

I like …'s strategy because …

Could you explain that in some other words?

Which strategy would be easier to scale up if more rows were added? Why?

## Create

Were there any questions asked at the start of the task that you now want to investigate?

Is there anything new that you are wondering now?

| | | |
|---|---|---|
| **rows** | **orientation** | **expand** |
| **change** | **multiply** | **pattern** |

## Consider

My tweet about the most important strategies for solving this problem is: …

I am not / fairly / very pleased with my work today because ...

To feel more positive about my work, I could adapt it by …

In the future, I might use ... because …

The resources that supported me were …

# Dice It Up

## Curious

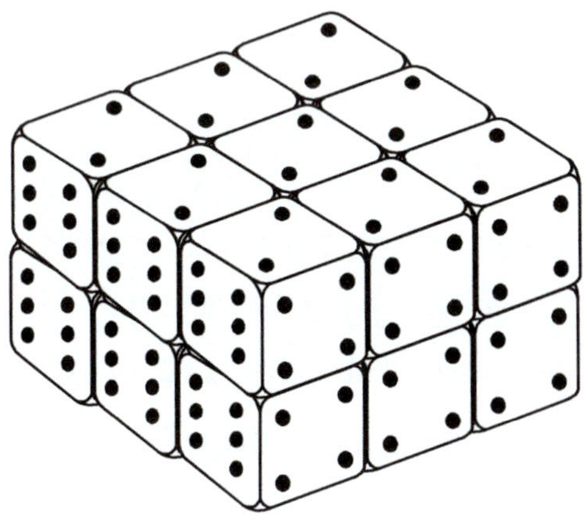

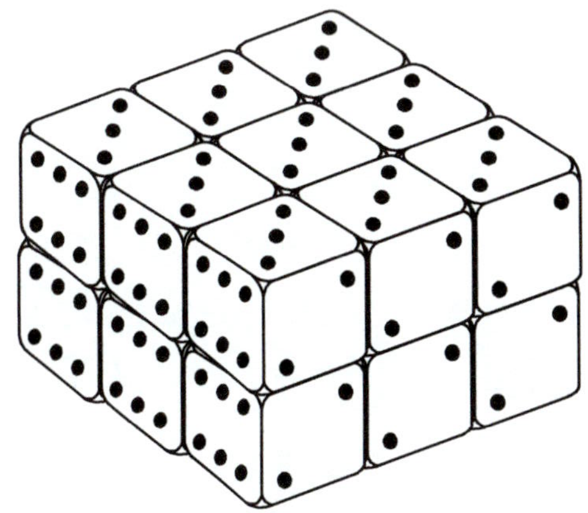

What do you notice?

What questions do you have?

**1** I can see …
I have noticed …
It looks like …

**2** I wonder what …
I wonder how many …
I wonder what would happen …

| | |
|---|---|
| **faces** | **visible** |
| **overall** | **total** |
| **opposite** | **multiple** |
| **interior** | **exterior** |

**3** My question to investigate is …

Curious

Express Pathway

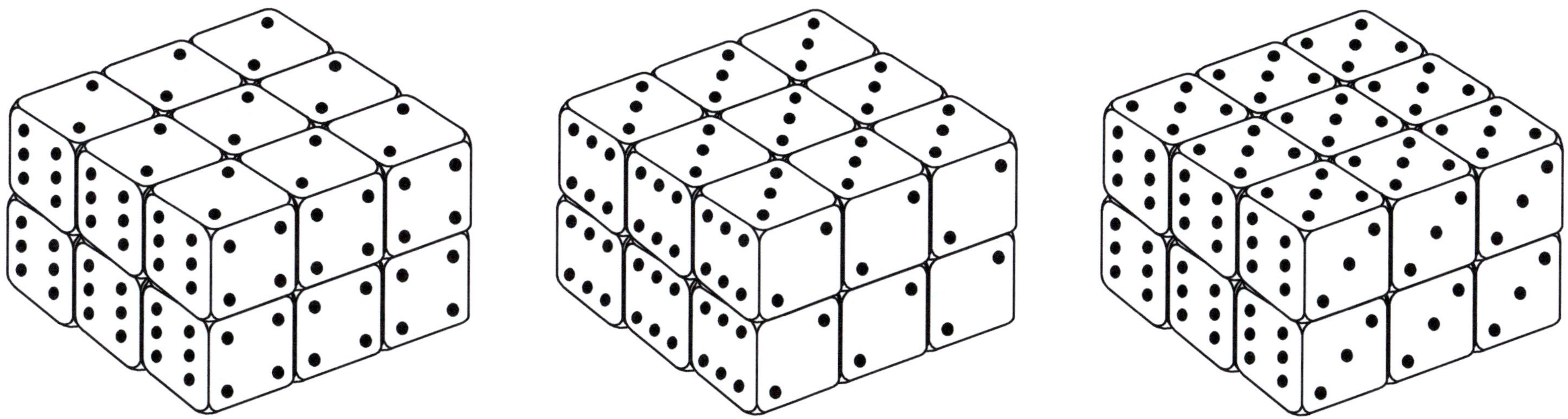

These cuboids built from dice have six faces.

The total number of dots around the outside of each cuboid is 147.

Can you explain why?

| | |
|---|---|
| faces | visible |
| overall | total |
| opposite | multiple |
| interior | exterior |

## Connect

### 1. Solve a simpler but related problem / make a model / act it out

Find the total for a one-layer cuboid first, and then a two-layer.

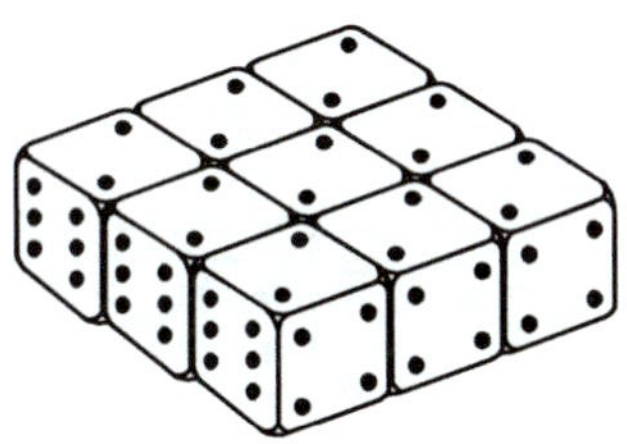

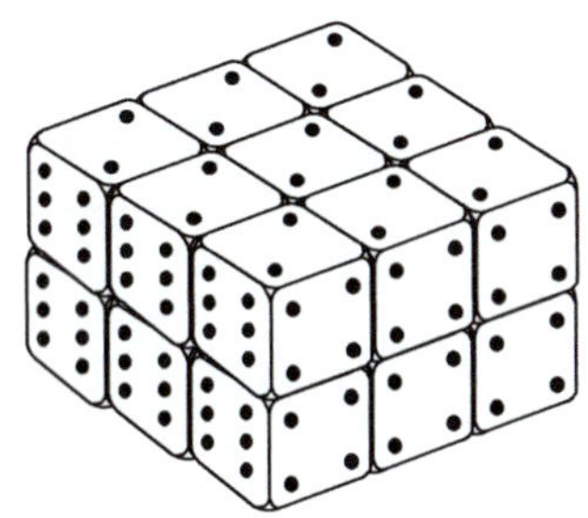

What's the same and what's different?

How might this help you solve the original problem?

### 2. Draw a diagram (1)

Fill in grids to show the number of dots on each face.

Visible faces

Hidden faces

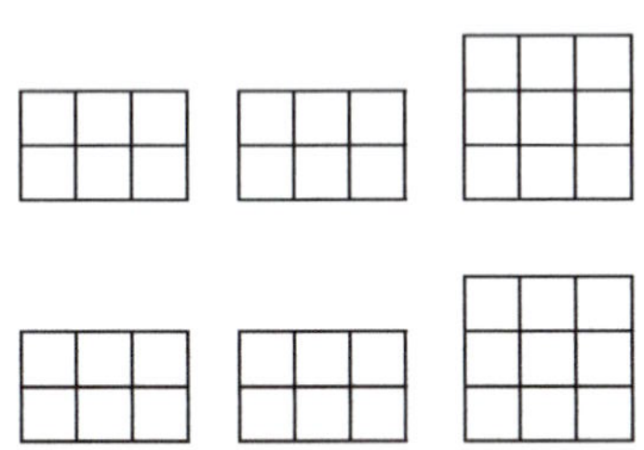

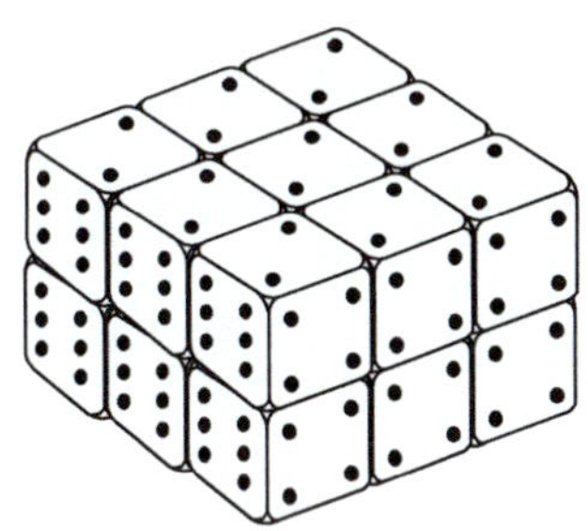

### 3. Draw a diagram (2)

How many dots in total on 18 individual dice?

How many dots are hidden when two faces touch in this cuboid?

How many pairs of faces touch?

Use 3 by 3 grids to help you work out the total number of hidden dots.

Subtract this from your original total.

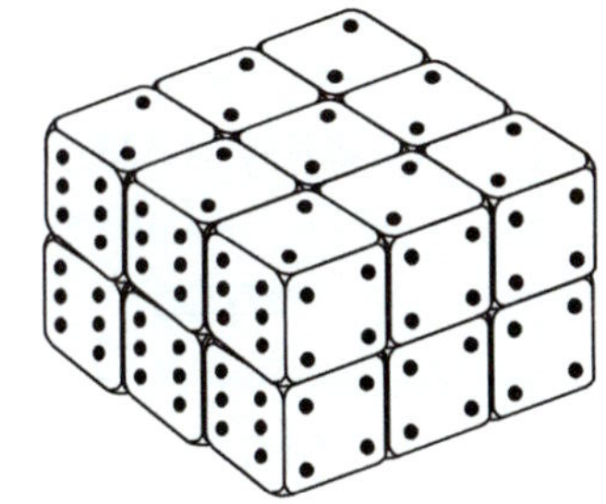

### 4. Make a table / spot a pattern

| Number of dots shown on dice | On visible faces | On hidden faces | Total number of dots |
|---|---|---|---|
| | | | |

Model 1

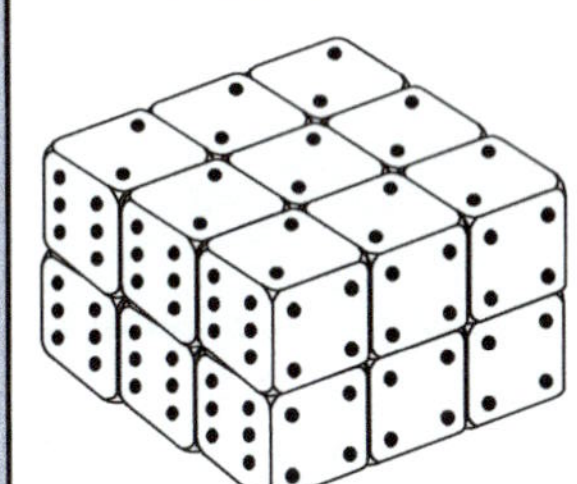

Model 2

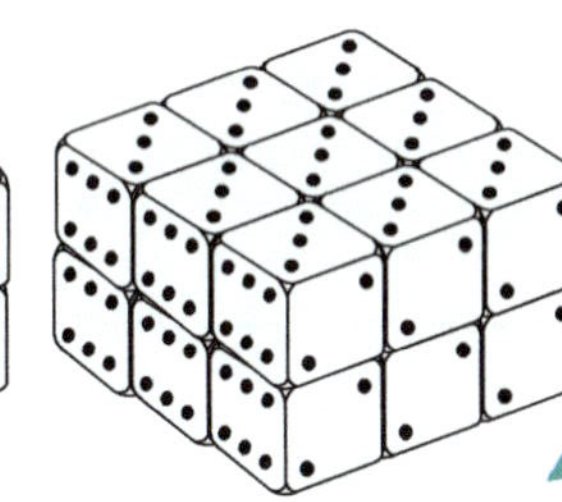

Complete a table for each model.

What patterns can you see in your completed tables?

I have noticed …
I predict that …

## Collaborate

Discuss your chosen strategy with the group.

What was your starting point?

Did others in the group have the same starting point or a different one?

## Communicate

Sit opposite a partner with a cuboid built from dice between you.

Use what you can see to make statements about what your partner can see.

- How many dots are visible to your partner?
- How do you know?

Score a point for each correct statement.

Then discuss the strategies you used in the **Connect** tasks to find the total number of dots. Use the word bank below.

| | | | |
|---|---|---|---|
| **visible** | **first** | **when** | **product** |
| **if** | **unknown** | **next** | **finally** |
| **multiple** | **because** | **as a result** | **therefore** |

## Create

Were there any questions asked at the start of the task that you now want to investigate?

Is there anything new that you are wondering now?

| | | |
|---|---|---|
| **more layers** | **3 by 3 by 3 cube** | **3 by 3 by 4 cuboid** |

## Consider

Two things that I found tricky in this task were …

I found it frustrating when … because …

I was surprised when I found out that ...

Working with others helped / did not help when …

# Missing Digit Multiplication

## Curious

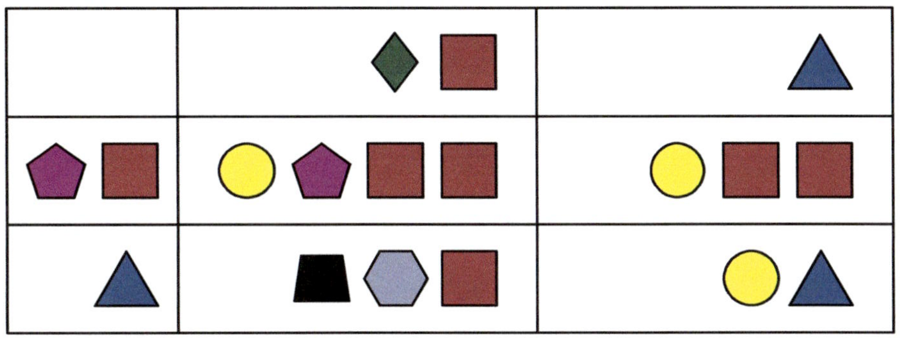

What do you notice?

What questions do you have?

**1** I can see …
I have noticed …
This reminds me of …

**2** I wonder if …
I wonder what …
I wonder why …
I wonder how …

| | | | |
|---|---|---|---|
| **multiplication** | **shape** | **represent** | **digit** |
| **start** | **solution** | **grid** | **value** |
| **product** | **partial product** | **multiplier** | **multiplicand** |

**3** My question to investigate is …

## Curious

## Express Pathway

Each shape in this formal written multiplication represents a different digit between 0 and 9.

Find what each shape represents.

How will you begin?

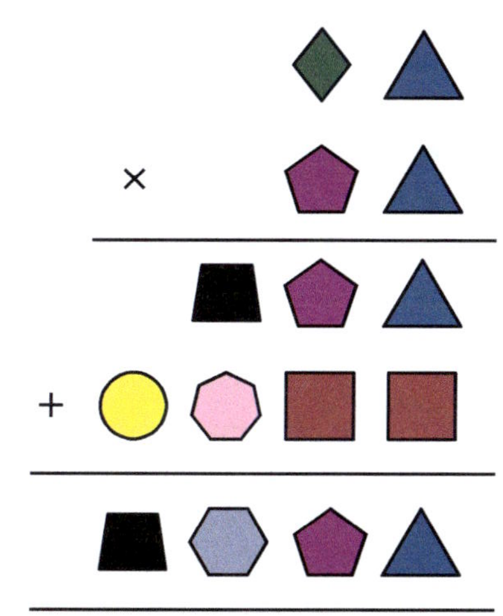

A useful starting point would be … because …

| | | | |
|---|---|---|---|
| multiplication | shape | represent | digit |
| start | solution | grid | value |
| product | partial product | multiplier | multiplicand |

## Connect

### 1. Solve a simpler but related problem

**Grid 1**

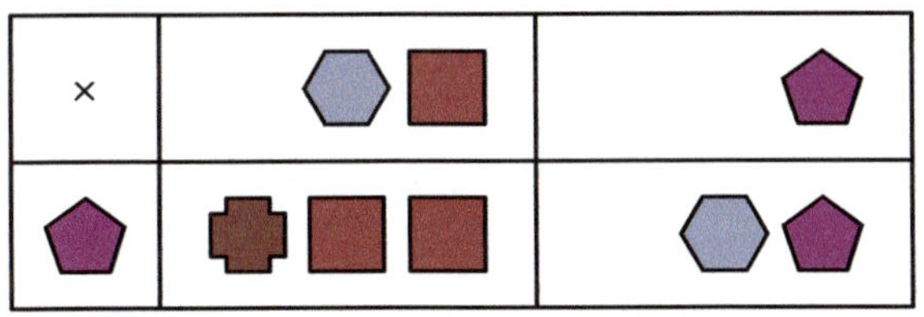

**Grid 2**

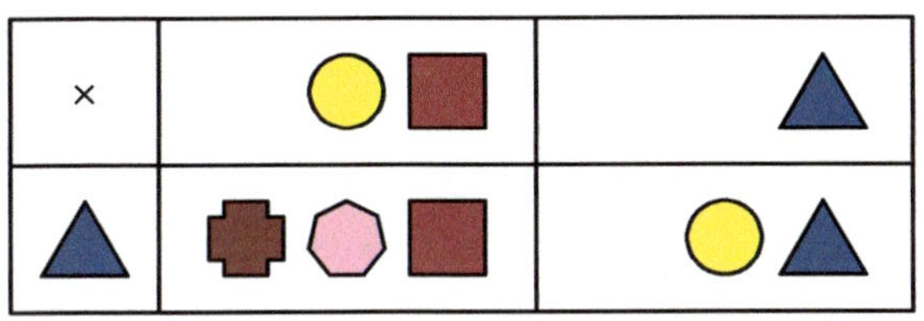

What's the same and what's different about each grid?

What do you notice about the shapes?

Where would be a good starting point?

### 2. Try then improve

I think that the … could be a …
because I have noticed …
I will try … for the … because …
… did not work for the … because …
Now that I know this, I will try …

### 3. Create a systematic list / make a table / spot a pattern / try then improve

How could you keep track of the digits that you have tried and either eliminated or kept?

Which rules do you need to ensure that the digits satisfy?

### 4. Make a table / spot a pattern

What observations can you make about each shape?

Does this help you to identify possible digits?

## Collaborate

Compare your ideas to someone else's.

Have you used a similar strategy?

## Communicate

Illustrate the process that you went through once you worked out the value of the red square.

Where did you start? What did you do next?

Could you turn this into an animation, video or presentation?

Make a plan using a storyboard and write a script for each section, using the conjunctions in the word bank.

| | | | |
|---|---|---|---|
| at first | before | because | consequently |
| next | eventually | due to this, | however |
| until | finally | therefore | then |
| meanwhile | concurrently | as a result | whereas |

## Create

Were there any questions asked at the start of the task that you now want to investigate?

Is there anything new that you are wondering now?

| | | |
|---|---|---|
| solvable | my own puzzle | other operations |

## Consider

I can connect the maths that I used in this task to …

The target I would set myself to improve the next time I try a similar problem is … I have chosen this as a target because …

I would give someone the following tips if solving a similar problem …

# The Perfect Number

## Curious

| | |
|---|---|
| 6 | 4   18   48<br>20   50   35 |

What do you notice?

What questions do you have?

**1** I can see …
I have noticed …
This reminds me of …

**2** I wonder why …
I wonder if …
I wonder what …

| | |
|---|---|
| **group** | **factors** |
| **prime numbers** | **factor sum** |
| **composite** | **perfect number** |
| **less than 50** | **systematic** |

**3** My question to investigate is …

**Curious**

**Express Pathway**

A perfect number is a number whose factor sum (excluding the number itself) is equal to the number.

16 is not a perfect number: $1 + 2 + 4 + 8 < 16$

39 is not a perfect number: $1 + 3 + 13 < 39$

Are there any perfect numbers between 0 and 50?

How do you know?

| | |
|---|---|
| **group** | **factors** |
| **prime numbers** | **factor sum** |
| **composite** | **perfect number** |
| **less than 50** | **systematic** |

## Connect

### 1. Solve a simpler but related problem

| | | | | | | | | | |
|---|---|---|---|---|---|---|---|---|---|
| 1 | 2 | 3 | 4 | 5 | 6 | 7 | 8 | 9 | 10 |
| 11 | 12 | 13 | 14 | 15 | 16 | 17 | 18 | 19 | 20 |
| 21 | 22 | 23 | 24 | 25 | 26 | 27 | 28 | 29 | 30 |

Why can't the numbers with squares around them be perfect numbers?

What other numbers could you rule out immediately?

Which numbers will you test?

### 2. Make a model / act it out

On a grid numbered 1–50, draw a circle around numbers if they are perfect and a square if they are not.

Are there any numbers that you can rule out immediately?

I can rule out … because …

I know that because …

### 3. Create a systematic list / make a table / spot a pattern

How can you make sure that you have considered and tested all numbers up to 50?

How will you work systematically to find all the perfect numbers between 1 and 50?

Where will you start?

### 4. Make a table / spot a pattern

How could you prioritise which numbers to test?

Can you rule out certain numbers? If so, why?

## Collaborate

How will you interact with other people to find perfect numbers?

Why would working as a group be useful?

Why would working on your own be useful?

## Communicate

Present a balanced explanation of why 6 and 28 are, or are not, perfect numbers.

Things to consider:

- What other numbers do you find interesting?
- What makes your chosen numbers interesting?
- What is a balanced argument?
- How will you structure your presentation?

> ***6 and 28 – the perfect numbers?***
>
> *Is it true that 6 and 28 are perfect numbers? If so, what is it that makes them perfect?*
>
> *Are there other numbers that (although not called 'perfect') are equally, or even more, interesting?*

| | | | |
|---|---|---|---|
| **… which** | **in addition, …** | **furthermore …** | **it is clear to see that …** |
| **many claim that …** | **moreover, …** | **we should also consider …** | **on the other hand, …** |
| **in contrast, …** | **on the contrary, …** | **in conclusion …** | **it is clear in my mind that …** |
| **factor** | **prime** | **composite** | **multiple** |

## Create

Were there any questions asked at the start of the task that you now want to investigate?

Is there anything new that you are wondering now?

| | | |
|---|---|---|
| **greater than 100** | **factor sum** | **larger / smaller than the number** |

## Consider

I would say that I am … out of 10 sure of my answer.

The most important mathematics I learned while working on this task was …

When I found a perfect number, I felt …

In order to find perfect numbers, I worked with others by …

I used … to help me. It was helpful because …

# Algebra is Amazing

## Curious

**A**

Think of a number.
Add 2.
Double it.
Subtract 2.
Divide by 2.
Subtract your original number.

Your answer is … **1**!

**B**

Think of a number.
Multiply by 2.
Add 4.
Halve it.
Subtract your original number.

Your answer is … **2**!

**C**

Think of a number.
Triple it.
Add 9.
Divide by 3.
Subtract your original number.

Your answer is … **3**!

What do you notice?

What questions do you have?

**1** I have noticed …
I can see …

**2** I wonder if …
I wonder whether …
I wonder how …

| | |
|---|---|
| **different** | **prove** |
| **instructions** | **algebra** |
| **four** | **answer** |
| **inverse** | **operation** |

**3** My question to investigate is …

## Curious

## Express Pathway

I can work out what number people are thinking of.
Here are some of my tricks.

**A**

Think of a number.
Add 2.
Double it.
Subtract 2.
Divide by 2.
Subtract your original number.

Your answer is … **1**!

**B**

Think of a number.
Multiply by 2.
Add 4.
Halve it.
Subtract your original number.

Your answer is … **2**!

**C**

Think of a number.
Triple it.
Add 9.
Divide by 3.
Subtract your original number.

Your answer is … **3**!

I want a trick to make people think of 4.

Can you write it for me?

| | | | |
|---|---|---|---|
| **different** | **instructions** | **four** | **inverse** |
| **prove** | **algebra** | **answer** | **operation** |

## Connect

### 1. Solve a simpler but related problem

Can you see pairs of instructions that cancel each other out?

Try different starting numbers for this trick.

Write down the answer at each step.

What do you notice?

> Think of a number.
> Double it.
> Add 2.
> Halve it.
> Subtract your original number.
> Your answer is … 1!

### 2. Make a model / act it out

Use number rods to make a model.

The first two steps of Trick A could be built like this.

| | |
|---|---|
| Think of a number. | 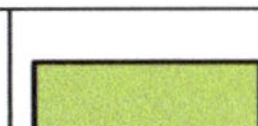 |
| Add 2. |  |

### 3. Make a table / spot a pattern

Complete this table for Trick A.

| Step | Number | Algebraic expression |
|---|---|---|
| Think of a number. | 4 | $n$ |
| Add 2. | 4 + 2 | $n + 2$ |
| Double it. | | |
| Subtract 2. | | |
| Divide by 2. | | |
| Subtract your original number. | | |

Can you make a table for the other tricks?

## Collaborate

Work with others to make a set of tricks that have the answers 1 to 10.

Think about disguising the language so it is not so easy to spot the inverses. Your aim is to amaze!

## Communicate

The stage is calling, get ready for your moment in the spotlight!
Choose your best 'Think of a number' trick and prepare to perform it to amaze your audience.

**Top tips for a live performance**

What are you going to call your act?

Is it a solo act, or a double act?

Think about your audience. Are your calculations achievable for them?

How can you add drama? What about a big reveal?

Do you need any props?

## Create

Were there any questions asked at the start of the task that you now want to investigate?

Is there anything new that you are wondering now?

| | |
|---|---|
| **other sorts of tricks** | **longest set of instructions** |
| **percentages or fractions** | **order** |

## Consider

I think that the proof that shows most clearly how the 'tricks' work …

I think that some children might still struggle to understand how the tricks work because … To help them I could …

I was surprised that … because …

I enjoyed / did not enjoy this task because …

To work more successfully in a group in the future, I could …

# Ancient Egyptian Fractions

## Curious

The Ancient Egyptians represented fractions differently to the way we do now.

| Our representation | Ancient Egyptian representation |
|---|---|
| $\frac{3}{10}$ | $\frac{1}{4} + \frac{1}{20}$ |
| $\frac{2}{3}$ | $\frac{1}{2} + \frac{1}{6}$ |
| $\frac{3}{5}$ | $\frac{1}{2} + \frac{1}{10}$ |

What do you notice?

What questions do you have?

**1** I can see …
I have noticed …

**2** I wonder how …
I wonder what …
I wonder if …

| | | |
|---|---|---|
| **unit fraction** | **numerator** | **Egyptian** |
| **fraction** | **equal** | **sum** |
| **three-quarters** | **three-eighths** | **five-ninths** |

**3** My question to investigate is …

**Curious**

## Express Pathway

In Ancient Egypt, fractions were written as a sum of different unit fractions.

All the denominators in the sum were different from each other.

Examples:

$\frac{1}{2}$ could have been expressed as $\frac{1}{3} + \frac{1}{6}$.

$\frac{1}{3}$ could have been expressed as $\frac{1}{4} + \frac{1}{12}$.

$\frac{5}{9}$ could have been expressed as $\frac{1}{2} + \frac{1}{18}$.

$\frac{7}{20}$ could have been expressed as $\frac{1}{4} + \frac{1}{10}$.

Express these fractions in the Ancient Egyptian format.

$\frac{3}{10}$ $\frac{2}{3}$ $\frac{3}{5}$

| | | |
|---|---|---|
| unit fraction | numerator | Egyptian |
| fraction | equal | sum |
| three-quarters | three-eighths | five-ninths |

## Connect

### 1. Draw a diagram / make a model / act it out

Convert $\frac{3}{4}$ into an Ancient Egyptian fraction.

Draw a diagram, or make a model using clay.

Think about if you had 3 tray bakes to share between 4 people.

Start by giving each person the largest unit fraction possible.

Can you give them $\frac{1}{2}$ each?

How much is left?

Share what is left between the 4 people. Use the largest unit fraction possible.

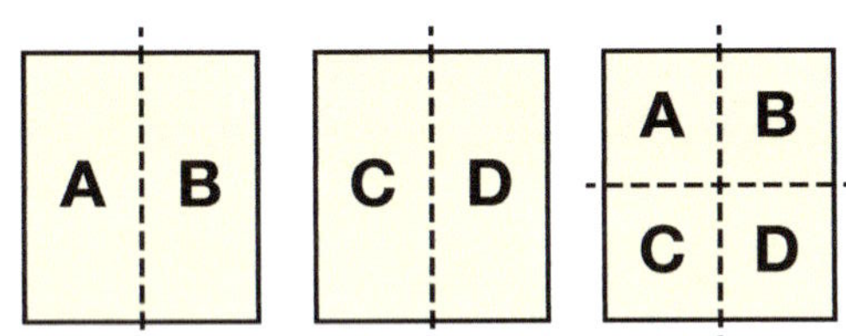


What does each person have?

Now draw diagrams or make models to convert $\frac{3}{8}$, $\frac{3}{10}$, $\frac{3}{5}$ and $\frac{2}{3}$ into Ancient Egyptian fractions.

### 2. Work backwards

Instead of adding up to a fraction, work backwards.

Subtract unit fractions from the fraction you wish to convert.

Start with the largest unit fraction you can. The largest unit fraction is $\frac{1}{2}$.

$\frac{3}{4} - \frac{1}{2} = ?$     $\frac{3}{4} = \frac{1}{2} + ?$

Repeat for $\frac{3}{8}$, $\frac{3}{10}$, $\frac{3}{5}$, $\frac{2}{3}$.

## Collaborate

As a group, write down the fractions you've worked with and their Ancient Egyptian format.

How can you organise these fractions? Can you order them by the size of the unit fractions in the Egyptian format?

Use what you now know to compare fractions $\frac{5}{8}$ and $\frac{6}{10}$.

## Communicate

Ancient Egyptians used hieroglyphics to write numbers. They represented fractions like this.

$\frac{1}{3}$

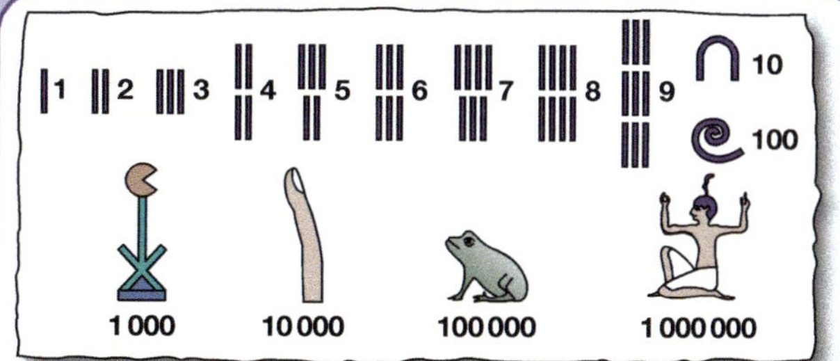


The top hieroglyph was a mouth. This indicated that it was a 'part' or fraction. The denominator used the symbols for whole numbers.

Create a 'Master code' sheet that shows the match between the hieroglyphics, the modern fraction and the 26 letters of the alphabet. For example:

| Letter | A | B | C |
|---|---|---|---|
| Modern fraction | $\frac{2}{5}$ | $\frac{5}{8}$ | $\frac{3}{4}$ |
| Ancient Egyptian fraction hieroglyphs |  |  |  |

Now write your secret message using Ancient Egyptian fraction hieroglyphics. Give your message to a partner to decode. They will need a 'Solve it' sheet, showing the alphabet and the matching fraction for each letter, but not the hieroglyphics.

## Create

The Rhind Mathematical Papyrus table was created by the Ancient Egyptians to make their calculations quicker. Here is an example.

| | | | | | | | |
|---|---|---|---|---|---|---|---|
| 5 | 3 | 15 | | 17 | 12 | 51 | 68 |
| 7 | 4 | 28 | | 19 | 12 | 76 | 114 |
| 9 | 6 | 18 | | 21 | 14 | 42 | |
| 11 | 6 | 66 | | 23 | 12 | 276 | |
| 13 | 8 | 52 | 104 | 25 | 15 | 75 | |
| 15 | 10 | 30 | | 27 | 18 | 54 | |

What do you think the table shows? How does it work?

Can you create a similar table?

## Consider

The trickiest part of my investigation was …

My thinking was influenced by …

I organised my work by …

Next time, I will …

# Puzzling Percentages

## Curious

7% of 300 = 300% of 7

What do you notice?

What questions do you have?

**1** I can see …
I have noticed …
This reminds me of …

**2** I wonder what …
I wonder if …
I wonder how …
I wonder whether …

| | | |
|---|---|---|
| prove | true | false |
| never true | always true | sometimes true |
| value | equal | representation |
| calculation | not a multiple of | greater than 100 |

**3** My question to investigate is …

**Curious**

**Express Pathway**

Is this equation always true, sometimes true or never true?

$$a\% \text{ of } b = b\% \text{ of } a$$

How will you use examples to prove or disprove this?

How will you make use of different manipulatives?

| | | |
|---|---|---|
| prove | true | false |
| never true | always true | sometimes true |
| value | equal | representation |
| calculation | not a multiple of | greater than 100 |

## Connect

### 1. Solve a simpler but related problem

What's the same and what's different in each pair of equations?

Use the first equation to help you solve the one below it.

- 50% of 100 = 100% of 50
  50% of 10 = 10% of 50
- 25% of 100 = 100% of 25
  25% of 60 = 60% of 25
- 10% of 100 = 100% of 10
  10% of 40 = 40% of 10

**Remember!**

50% is equivalent to $\frac{1}{2}$

10% is equivalent to $\frac{1}{10}$

25% is equivalent to $\frac{1}{4}$

60% is 50% + 10% or $\frac{6}{10}$

10% is equivalent to $\frac{1}{10}$

40% is 50% – 10% or 4 lots of 10% or $\frac{4}{10}$

### 2. Make a model / act it out

How will you build each side of the equation?

How will you compare your models?

Which concrete resources do you think are most useful?

### 3. Draw a diagram

What will your bar models look like?

What do you need to consider?

How clear do you think your bar models are?

What could you do to improve them?

## Collaborate

Choose an equation and a type of concrete resource.

Prepare your proof.

Share it with the rest of your group.

Listen carefully to the explanations of your group.

How clear are the explanations?

I'm sorry, I don't understand. Could you explain that in a different way?

## Communicate

Create a presentation which involves using manipulatives to prove the equality of the equations you have explored.

You could choose one equation and prove this in a variety of ways or choose one manipulative to prove a variety of equations.

Make sure to consider the following points:

For a filmed presentation:

- Would preparing a script be useful?
- How will the camera clearly show the manipulatives?
- How will you engage your audience?

For a live presentation:

- Would preparing note cards be useful?
- How will you clearly model to your audience?
- How will you engage your audience?

## Create

Were there any questions asked at the start of the task that you now want to investigate?

Is there anything new that you are wondering now?

| | | |
|---|---|---|
| **fractions** | **decimals** | **other relationships** |

## Consider

I think the representation that shows the equality clearest is …

Some children may still struggle to understand the equality because … To help them I would …

I was surprised about … because …

# The Red Rectangle

## Curious

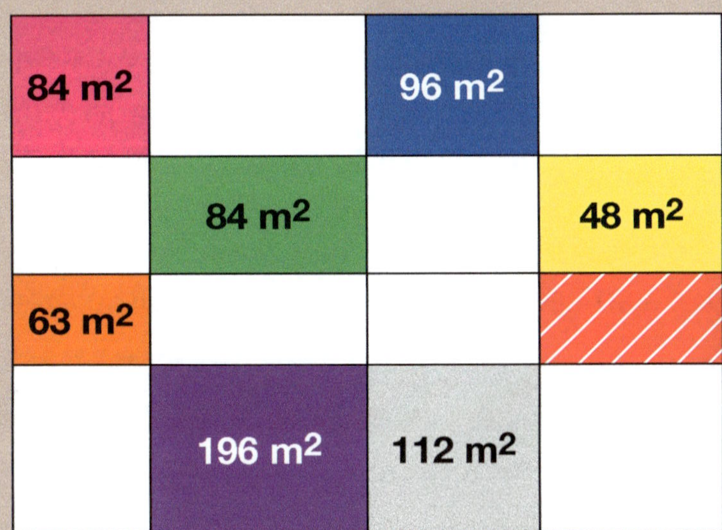


Note:

The diagram is not drawn to scale.

What do you notice?

What questions do you have?

**1** I wonder …
I can see …
I have noticed …
This reminds me of …
I know that …

**2** I wonder whether …
I wonder what …
I wonder if …

| | | |
|---|---|---|
| **length** | **common factor** | **square number** |
| **multiple** | **area** | **dimension** |
| **factor** | **row** | **column** |

**3** My question to investigate is …

## Curious

## Express Pathway

Calculate the area of the red striped rectangle.

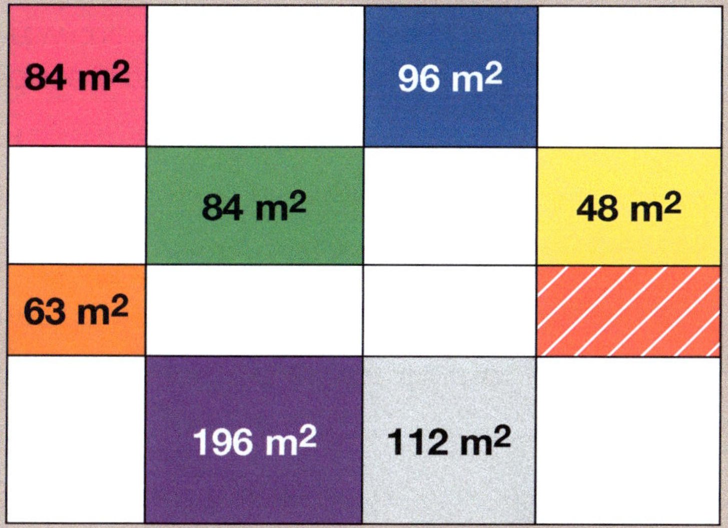


Note:

The diagram is not drawn to scale.

| | | |
|---|---|---|
| length | common factor | square number |
| multiple | area | dimension |
| factor | row | column |

## Connect

### 1. Solve a simpler but related problem

Each shaded rectangle has whole-number dimensions between 1 m and 10 m.

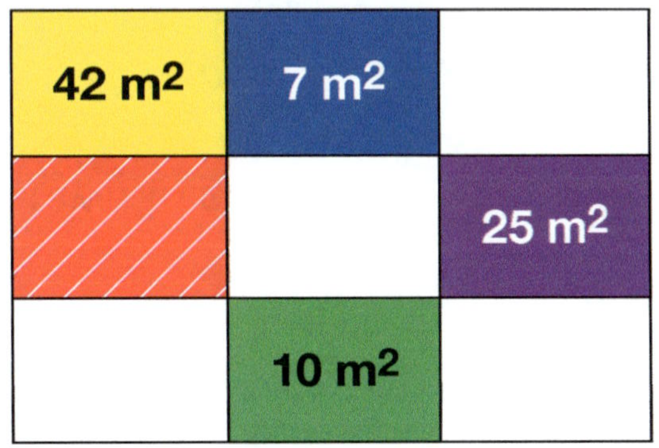


Find the area of the red striped rectangle.

Where will you start?

What do you know about the number …?

What do you know about the factors of …?

What clues does that give you? What will you do next?

### 2. Try then improve

Choose a rectangle to start with.

Which factor pair will you try first?

How will you keep track of what you have already tried?

If the area is … then …
If I know this, then I also know …
It will help to look for …
The most efficient starting place will be …

### 3. Draw a diagram / create a systematic list

How could you draw a diagram to help you to organise your thoughts?

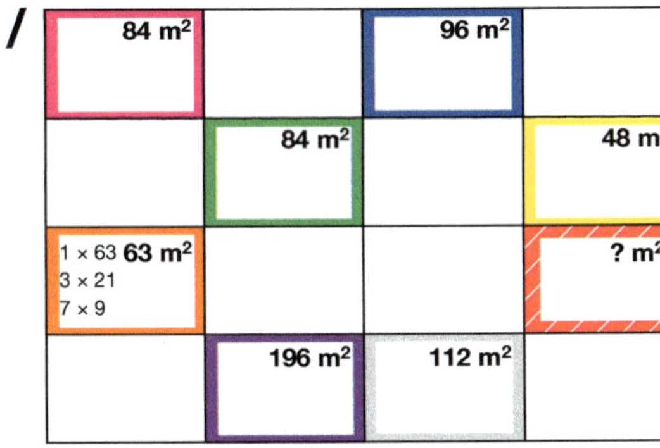


### 4. Make a table / spot a pattern

How might a table help you see the possible dimensions more clearly?

When you have made your table, can you start testing possible dimensions?

Which factors can you disregard?

What is the relationship between the rows and columns?

The …. column can't be … because …
The … row can't be … because …
If … doesn't work then I also know ….

## Collaborate

How was working in a pair helpful for this task?

## Communicate

How could you explain to someone how you found the area of the red striped rectangle?

How could you use a flow chart to make your reasoning clearer?

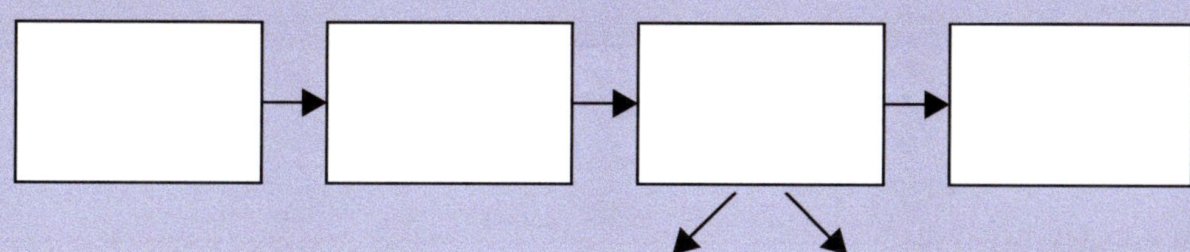

| | | |
|---|---|---|
| **column** | **row** | **given area** |
| **factor** | **factor pair** | **common** |
| **possibilities** | **the remaining factors ...** | **could (not) be ...** |
| **doesn't work** | **certain that ...** | **Now that we know this ...** |
| **Once we deduced this ...** | **This meant that ...** | **As a result ...** |

## Create

Were there any questions asked at the start of the task that you now want to investigate?

Is there anything new that you are wondering now?

| | | |
|---|---|---|
| **remove one area label** | **my own missing-area puzzle** | **larger grid** |

## Consider

The trickiest part of the problem was … I think that this was because …

I was frustrated by … I think that was because … To overcome this, I …

I said 'aha' when … As a result, I …

… influenced my thinking because … This changed / did not change my decisions because …

# The Tale of Two Squares

## Curious

$2^2 - 1^2 = 2 + 1$

$3^2 - 2^2 = 3 + 2$

$4^2 - 3^2 = 4 + 3$

What do you notice?

What questions do you have?

**1** I can see …
I have noticed …

**2** I wonder what …
I wonder if …
I wonder whether …

| | |
|---|---|
| difference | consecutive |
| square | pattern |
| adding | between |
| pair | predict |

**3** My question to investigate is …

**Curious**

**Express Pathway**

You can find the difference between the squares of two consecutive numbers by adding the numbers together.

Is this statement true or false?

Example: $2^2 - 1^2 = 2 + 1$

Do you notice any patterns during your investigation?

| | |
|---|---|
| **difference** | **consecutive** |
| **square** | **pattern** |
| **adding** | **between** |
| **pair** | **predict** |

## Connect

### 1. Make a model / act it out

Can you make models to show that these equations are true?

**$2^2 - 1^2 = 2 + 1$**

**$3^2 - 2^2 = 3 + 2$**

**$4^2 - 3^2 = 4 + 3$**

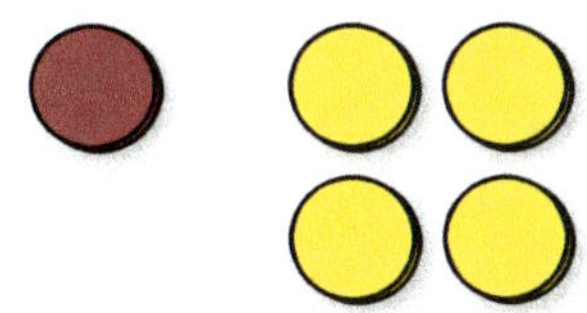

What concrete resources will you use to make your models?

### 2. Draw a diagram

How could you draw $2^2 - 1^2$?

Is the difference equal to 2 + 1?

What do you notice?

Is this similar for $3^2 - 2^2$?

What do you notice now?

Can you explain why?

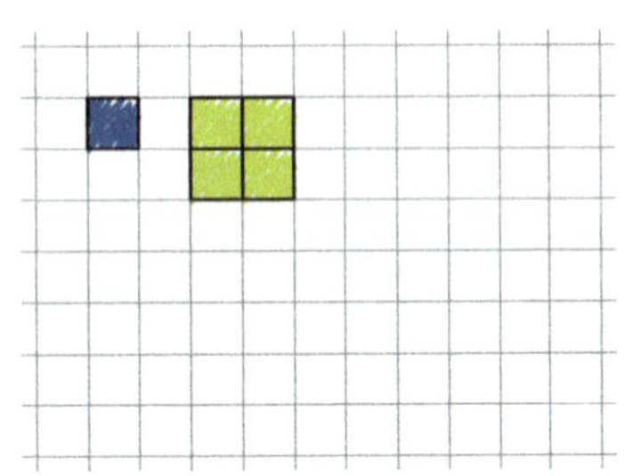

### 3. Make a table / spot a pattern

Could a table help you to see the patterns more easily?

What headings will you use?

| | | | |
|---|---|---|---|
| | | | |
| | | | |
| | | | |

Compare each row with the one below.

What is the same and what is different?

How can the equations in one row help you to complete the equations in the next row?

How can you prove what you have noticed?

## Collaborate

Compare and contrast your model or diagram with someone else's.

What do you notice?

## Communicate

Write a 'Did you know' interview for a magazine, explaining your findings.

> **Did you know?**
> ***A tale of two squares***
>
> *In this week's 'Did you know?' article, we interviewed . . . from . . . School, as a recent discovery has been brought to our attention.*
>
> *Did you know that . . .?*
>
> *Read on to discover more.*

| | | | |
|---|---|---|---|
| **… told us that …** | **"…," reported …** | **After that …** | **When …** |
| **At this point …** | **If … then …** | **Since …** | **It was then that …** |
| **You might be surprised to know that …** | **Now that you know more about …** | **This representation proves …** | **square** |
| **sum** | **difference** | **consecutive** | **non-consecutive** |

## Create

Were there any questions asked at the start of the task that you now want to investigate?

Is there anything new that you are wondering now?

| | | |
|---|---|---|
| **difference of 2** | **1–20** | **total** |
| **consecutive** | **patterns** | **cubed** |

## Consider

I am not / fairly / very confident about the proof that I have used because …

I think that my proof would / would not be clear enough to convince someone else what I have found out, because …

During my investigation, I said 'aha' because … As a result, I …

… influenced my thinking because … This changed / did not change my decisions because …

# Talented Tessellation

## Curious

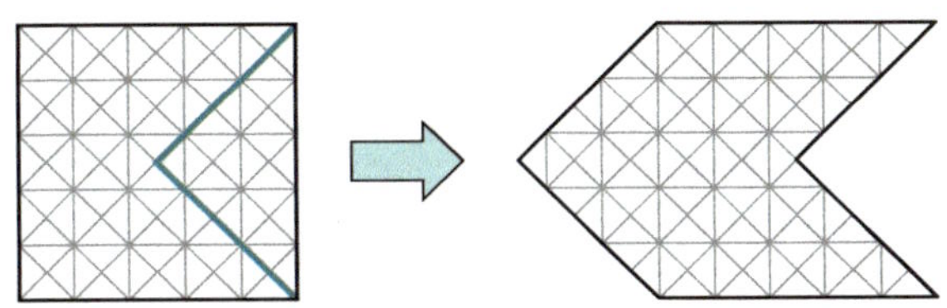

A

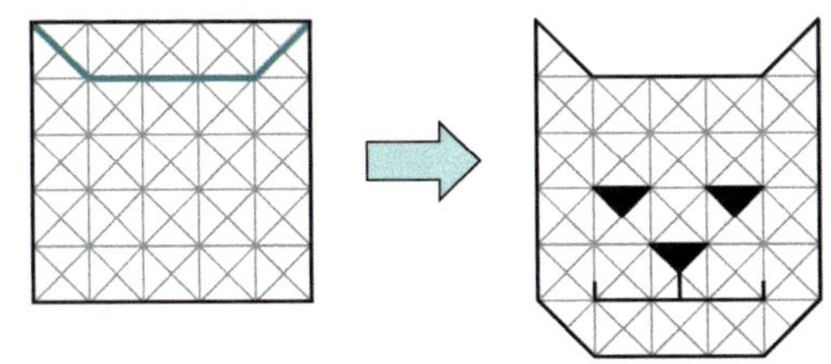

B

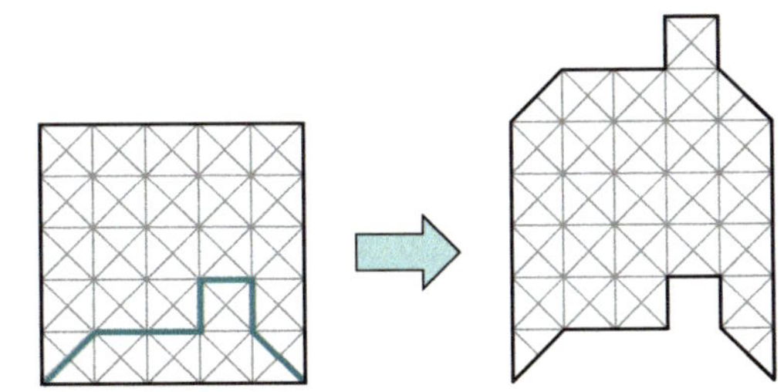

C

What do you notice?

What questions do you have?

**1** I can see …
I have noticed …
This reminds me of …

**2** I wonder if …
I wonder how …
I wonder whether …

| | | |
|---|---|---|
| **perimeter** | **area** | **centimetre (cm)** |
| **number of sides** | **tessellate** | **translate** |
| **centimetre squared ($cm^2$)** | **consistent** | **different** |

**3** My question to investigate is …

## Curious

## Express Pathway

Rosa, Rashid and Ren are making tessellating patterns.

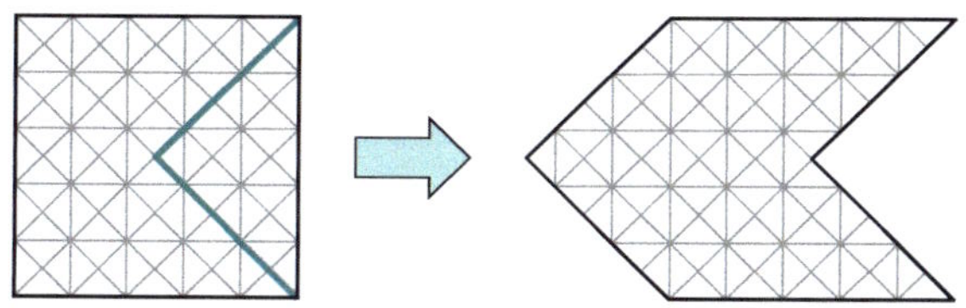

A

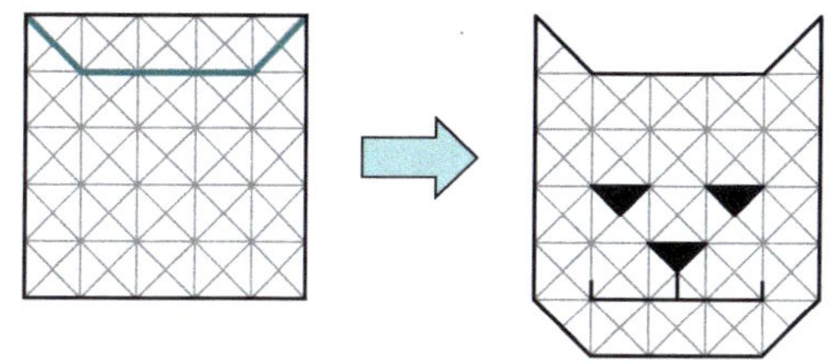

B

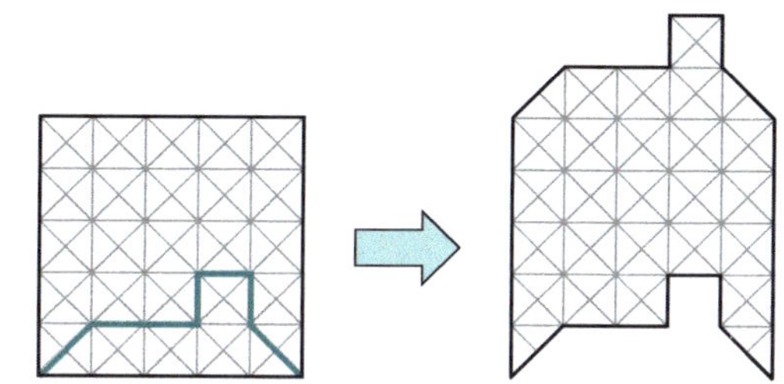

C

**Rosa** says: As the perimeter increases, then the area must also.

**Rashid** says: As you increase the number of sides on the tessellating shape, both the perimeter and the area will increase.

**Ren** says: There is a limit on how big both the perimeter and area can get.

Create some of your own tessellating shapes using 5 cm by 5 cm squares.

Who do you agree with?

Who do you disagree with?

Why?

| | | |
|---|---|---|
| **perimeter** | **area** | **centimetre (cm)** |
| **number of sides** | **tessellate** | **translate** |
| **centimetre squared ($cm^2$)** | **consistent** | **different** |

## Connect

### 1. Make a model / act it out / try then improve

Try making some more shapes that will tessellate, using the same 5 cm by 5 cm grid.

Measure the perimeter of each shape or use this information:

- each square is 1 cm by 1 cm;
- a diagonal is 1·4 cm.

Each time you try a new shape, can you build on what you have already tried, rather than starting from scratch?

### 2. Create a systematic list / make a model / act it out

How will you record the shapes you make, and the results?

What information will you record for each shape?

Can you spot any patterns in your notes?

Could you order your notes in a systematic way?

> *Shape A*
> *Sides: 6*
> *Perimeter: 24 cm*
> *Area: 25 $cm^2$*

### 3. Make a table / spot a pattern

Would a table help?

What will the headings of your table be?

What variables may be changing?

> I have added … more sides, so I predict the perimeter will increase by …
> I have changed …, so I predict that …
> In order to make the perimeter increase by … cm, I need to add … sides.

## Collaborate

In a small group, pool all of your tessellating shapes.

How many different ways can you order the information?

Can you work together to spot any new patterns?

## Communicate

Select a topic to research.

Prepare a written or verbal presentation.

- Rotational tessellation
- Reflective tessellation
- Regular and semi-regular tessellation
- Tessellation and art
- Tessellation and history

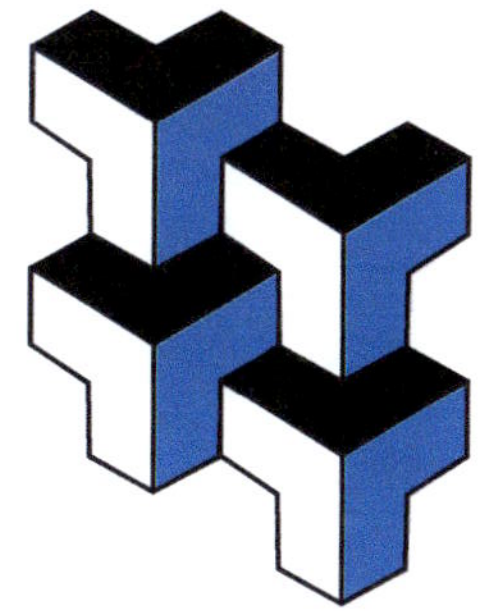

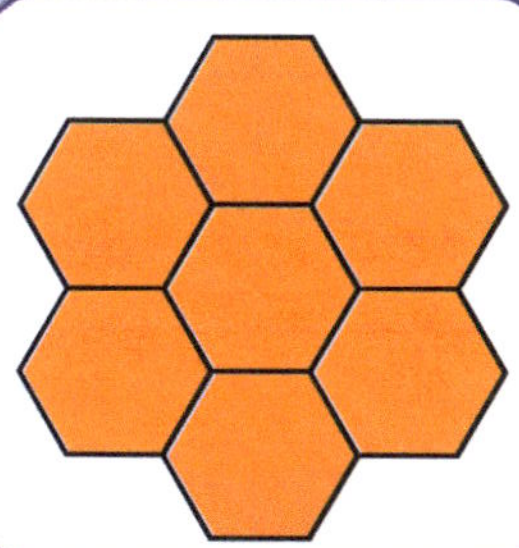

## Create

Were there any questions asked at the start of the task that you now want to investigate?

Is there anything new that you are wondering now?

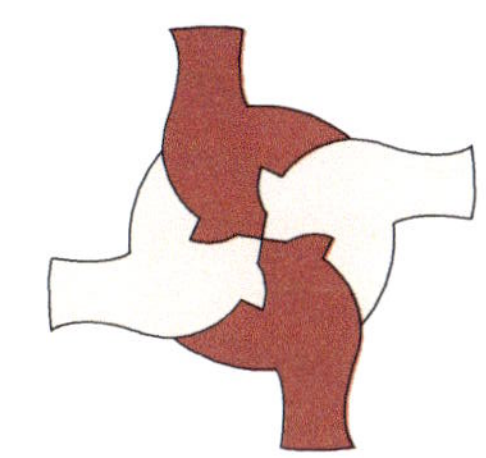

| | |
|---|---|
| **rotational** | **reflective** |
| **adjacent** | **opposite** |
| **more than one side** | **curved** |

## Consider

The most important mathematics I learned while answering this question was … It was important because …

I think some people might still struggle to understand the answer because …

The fun bit of my learning was … It was fun because …

I interacted with others by … This helped me …

# Battle of the Nets

## Curious

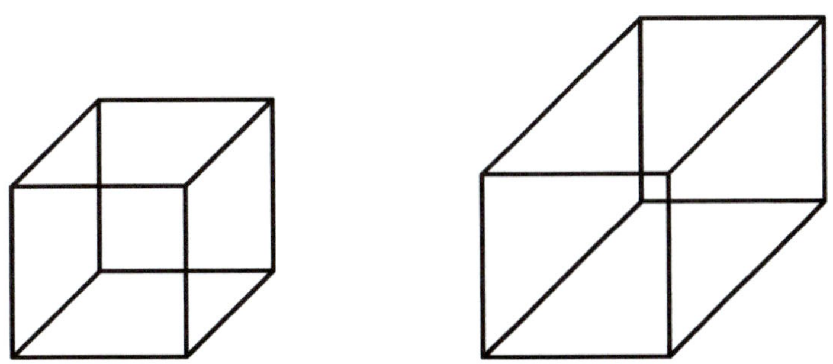

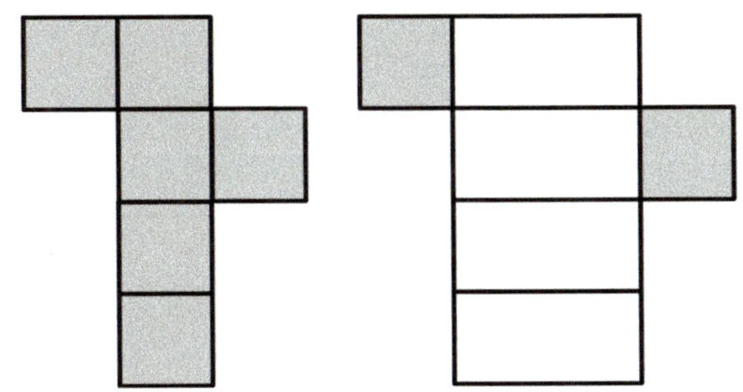

What do you notice?

What questions do you have?

**1** I can see …
I have noticed …

**2** I wonder if …
I wonder what …
I wonder how …
I wonder how many …

| | | |
|---|---|---|
| **net** | **cube** | **square cuboid** |
| **11 nets** | **less than** | **more than** |
| **track** | **all possibilities** | **systematic** |

**3** My question to investigate is …

Curious

## Express Pathway

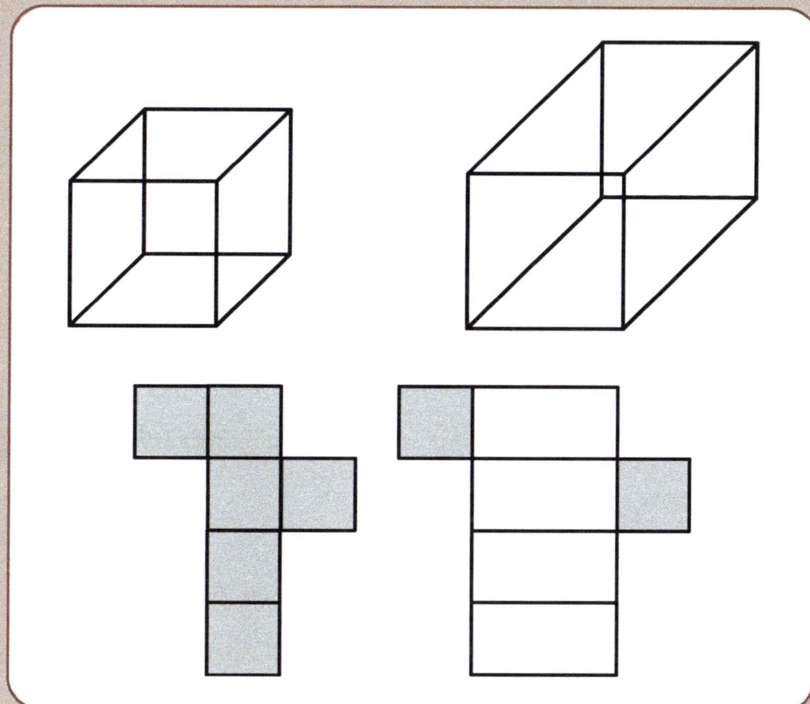

Here is a cube and a square cuboid.

(A square cuboid has two opposite square faces.)

Both of these shapes can be made from 11 different nets.

True or false?

| | | |
|---|---|---|
| **net** | **cube** | **square cuboid** |
| **11 nets** | **less than** | **more than** |
| **track** | **all possibilities** | **systematic** |

## Connect

### 1. Make a model / act it out

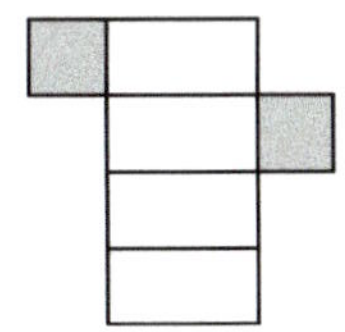

Make a net like this one.

Use the net to make a model of the shape.

What do you need to consider?

What will you use to ensure accuracy?

What dimensions will you use?

### 2. Try then improve

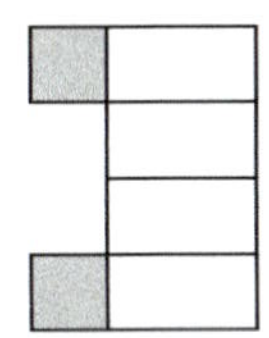

What will you do if a net does not work?

What have you learned from this attempt that could help you?

> When I folded up my net, I found that …
> Because of this, I moved …
> This helped because …

### 3. Draw a diagram

How will you accurately record your nets?

### 4. Create a systematic list

Can you create a list by drawing diagrams?

Does a list have to be in the form of a table?

## Collaborate

Compare and contrast the nets that you have made with someone else.

Have they found any that you have not?

How can you check that they are definitely 'new' nets, rather than rotations or reflections?

Have they been tested to check that they do form a square cuboid?

## Communicate

You have been commissioned by a games company to design and make a card game that will help children to rehearse their knowledge of nets of 3D shapes.

You should consider:

- your target audience;
- the rules of the game;
- resources that players will need;
- specific mathematical language;
- precise mathematical drawings.

**[*Game name*]**

Suitable for age … and up.
Number of players: …

The aim of the game is …

You will need:
…

How to play:
…

## Create

Were there any questions asked at the start of the task that you now want to investigate?

Is there anything new that you are wondering now?

| | | |
|---|---|---|
| **other 3D shapes** | **non-square faces** | **larger than the number** |

## Consider

Two things that were tricky in this task were … As a result, I felt … As a result, I …

I am not / fairly / very confident about what I found out as a result of my investigation because …

I was surprised by … because …

Working with others helped me because … But I felt hindered when …

# Distorted Data

Curious

## Facts presented by the Football Everywhere campaign

**Fact 1:**

95% of the children we asked love football.

Only 20 children were asked.
All of them are in the Y6 football team.

**Fact 2:**

The vast majority of the children we asked play football at playtime.

**Do you ever play football at playtime?**

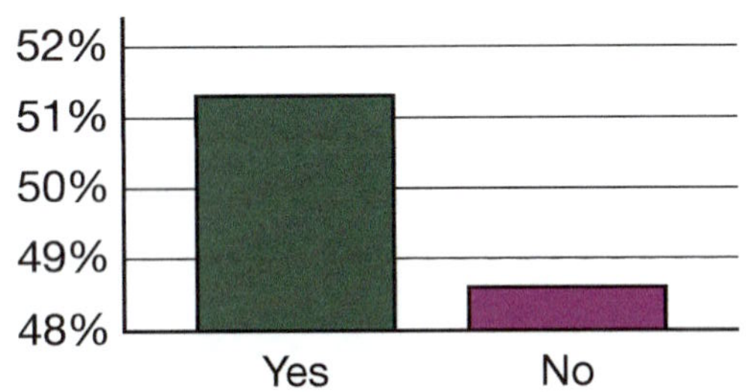


This is the data that was collected:

**Do you ever play football at playtime?**

| | |
|---|---|
| Yes | 19 |
| No | 18 |

**Fact 3:**

We presented three options for changes at playtime, and 9 out of 10 children chose football-only playtimes.

| Playtime options | | | | | | | | | |
|---|---|---|---|---|---|---|---|---|---|
| Football | ⚽ | ⚽ | ⚽ | ⚽ | ⚽ | ⚽ | ⚽ | ⚽ | ⚽ |
| Option 2 | | | | | | | | | |
| Option 3 | ⚽ | | | | | | | | |

60 children were asked.
The options were:
1) Football-only playtimes
2) Only silent and solitary playtimes
3) All playtimes cancelled.

**1** I can see …
I have noticed …
It reminds me of …

**2** I wonder if …
I wonder what …
I wonder how …

| | | | |
|---|---|---|---|
| **data** | **information** | **axis** | **opinion** |
| **distorted** | **misleading** | **true** | **false** |
| **chosen** | **persuade** | **present** | **purpose** |
| **option** | **sample** | **selective** | **bias** |

**3** My question to investigate is …

## Curious

## Express Pathway

### Facts presented by the Football Everywhere campaign

The campaign is led by the Year 6 football team. They have collected data to persuade the school that:

- all playground space should be used as football pitches;
- football should be the only activity during playtime.

95% of the children we asked love football.

Only 20 children were asked.
All of them are in the Y6 football team.

The vast majority of the children we asked play football at playtime.

**Do you ever play football at playtime?**

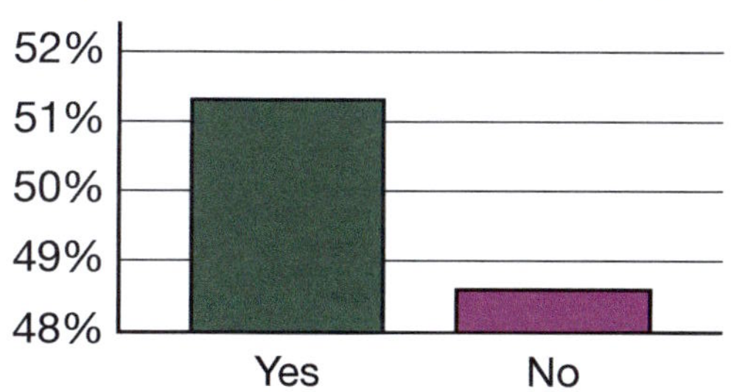


This is the data that was collected:

**Do you ever play football at playtime?**

| | |
|---|---|
| Yes | 19 |
| No | 18 |

We presented three options for changes at playtime, and 9 out of 10 children chose football-only playtimes.

| Playtime options | | | | | | | | | |
|---|---|---|---|---|---|---|---|---|---|
| Football | ⚽ | ⚽ | ⚽ | ⚽ | ⚽ | ⚽ | ⚽ | ⚽ | ⚽ |
| Option 2 | | | | | | | | | |
| Option 3 | ⚽ | | | | | | | | |

60 children were asked.
The options were:
1) Football-only playtimes
2) Only silent and solitary playtimes
3) All playtimes cancelled.

The three facts are accurate. Are you persuaded by them?

The children have been very clever in how they collected and presented the data. Can you see how?

What change would you like to campaign for in your school?

How could you collect data so it favours your argument?

How could you present the data so that it looks like everyone agrees with you?

You cannot make up the data. You must collect it from people in your school.

Your data must be accurate, but you can present it however you choose.

| | | | |
|---|---|---|---|
| data | information | axis | opinion |
| distorted | misleading | true | false |
| chosen | persuade | present | purpose |
| option | sample | selective | bias |

## Connect

### 1. Work backwards

The Football Everywhere campaign asked the question: ‘Do you love football?’ But they only asked children in the football team.

Think of a question you could ask children in your class.

- What answer do you want?
- Who is most likely to give you that response?

If you asked the whole class the same question, do you think the results would be different?

### 2. Try then improve

The Football Everywhere campaign asked children to say how they would like playtimes to change, but they only gave them three options: (1) football-only playtimes; (2) only silent and solitary playtimes; (3) all playtimes cancelled.

Think of a question you could ask children in your class.

- What answer do you want?
- How could you word the question to lead people towards that answer?

Test your question out with a small group.

Improve the wording of your question, then try again.

### 3. Draw a diagram

How could you present your data in a way that makes it seem to back up your argument more strongly?

Think about:

| | | |
|---|---|---|
| tables | title | percentages |
| graph type | labels | fractions |
| axis scale | | |

## Collaborate

Work together to agree:

- what your campaign aims to achieve;
- the best way to collect your data.

**[Campaign name]**

*We want to persuade our school to agree to: . . .*

*We believe this change would be good because . . .*

*We think we will get the result we want by collecting data from . . .*

*We will word our question like this: . . .*

## Communicate

Construct your persuasive argument using your facts and graphs and present it to another group or class.

Remember: you are not allowed to make facts up, but you can present them in a misleading way.

| | | | |
|---|---|---|---|
| **The fact is …** | **It is clear to see …** | **Obviously …** | **I firmly believe …** |
| **Everyone knows that …** | **…. strongly suggests that …** | **I'm sure that you would agree that …** | **Without a doubt …** |
| **Surely you must realise …** | **It must be agreed that …** | **Although …** | **However, …** |
| **On the other hand, …** | **It must also be considered …** | **Therefore …** | **In conclusion …** |

Consider another group's presentation.

- Have they told the whole story through their argument?
- What might they have modified or omitted in order to mislead their audience?
- What questions do you need to ask to discover the full truth?

## Create

Were there any questions asked at the start of the task that you now want to investigate?

Is there anything new that you are wondering now?

| | | |
|---|---|---|
| **real life** | **strategy** | **influence** |
| **honest** | **dishonest** | **unbiased** |
| **analyse** | **evaluate** | **assess** |

## Consider

I learned … during this task. I could use this in the future when …

The assumptions that I made at the beginning of the task turned out to be right / wrong because …

I was surprised when …

Out of 10, I would give myself a … for how well I worked with others, because …